Bibliographic information published by the German National Library:

The German National Library lists this publication in the National Bibliography; detailed bibliographic data are available on the Internet at http://dnb.dnb.de .

Imprint:

Print and binding: Books on Demand GmbH, Norderstedt Germany
ISBN: 9783346128171

This book at GRIN:

https://www.grin.com/document/519899

Mitku Alemu Mengistu

Impacts of Livestock Holdings on Households' Participation in Wood Fuel Production From Dryland Forest

Case study in Dire Dawa, Ethiopia

GRIN Verlag

Impacts of Livestock Holdings on Households' Participation in Wood Fuel Production from Dryland Forest (Case study in Dire Dawa; Ethiopia)

Mitku Alemu[1]*

Abstract

This study examines the influence of livestock holding on households' participation in wood fuel production from dryland forest. Survey data were collected from 320 households in Dire Dawa administration council of Oromia state, Ethiopia following a snowball sampling technique. Binary logit regression model and descriptive statistics were used to analyze the impacts of livestock holding on households' participation in fuelwood production from dryland forest. The estimation of the model indicates that livestock and size of arable land holdings were significant factors in determining households' participation in wood fuel production. An increase of one unit in livestock holding leads to the probability of participation in wood fuel production of households decreased by 0.118 units while the land size has a negative impact on the participation of wood fuel production. When land size increased by 1 hectare the probability of participating in wood fuel production will be decreased by 0.9 units. The mean of wood fuel producers' livestock holding was about 5.84 TLU and 10.67 for non- wood fuel producer which indicates the negative correlation of livestock holding and wood fuel production. Thus the study concludes that households with large livestock less likely to participate in wood fuel extraction from dryland forest. Promoting sustainable livestock production with appropriate grazing plan can have a significant role in dryland forest conservation and sustainable forest management.

Key word: - Livestock, dryland forest, wood fuel, households, forest management

Table of Contents

1. Introduction

Despite numerous importance of livestock production, it can cause habitat destruction and structural change of vegetation with many undesirable consequences especially in dryland forest and woodland ecosystems if sustainable management of dryland forest and woodland not implemented [1]. On the other hand, the presence of livestock in dryland forest has an important function in reducing forest fire by removing dry perennial plant material which aggravates forest fire. Study with livestock systematic grazing used as a treatment for combating desertification that indicating livestock herding and pastoralism culture has a positive impact on desertification reduction[2].

The objectives of this study are to identify the impacts of household livestock holdings in wood fuel production from dryland forest in the study area. This study will give a clearer picture of wood fuel producers' characteristics as well as the role of livestock holdings in determining households in wood fuel production participation. In order to prepare and implement dryland forest management deep knowledge on the relationship between livestock and forest extraction play a significant role in one hand and there is no strong evidence on livestock holdings impact on dryland forest in the study area on the other hand. To fill those gaps conducting this study is so urgent and critical. Mainly three livelihood strategies were employed in and around dryland forest ecosystem pastoralism agro-pastoralism and inactive agriculture with wood fuel production, mining and wild fruit collection also practiced in dryland areas of Ethiopia[3]. The dependency of the local community on dry land forest differs with the season, geographical position and socioeconomic condition of households[1]. Livestock production in dryland forest has both negative and positive impact on forest regeneration and development. Overgrazing is one major cause of dryland forest degradation especially in areas of pastoralist and agro-pastoralist community. The community in the study area can be classified into pastoralist and agro-pastoralist livelihood. Overgrazing imposes several interrelated problems in dryland forest ranging from browsing damage of seedling to the soil erosion associated with physical damage to the soil. About 90 % of agricultural land in Africa composed of rangelands which include woodlands sustaining more than 5 million people livelihoods[4]. Livestock production carried out in developing countries for several products and uses such as milk, meat, hides, skins, dung for energy sources and mainly to use oxen for ploughing especially in Ethiopian rural areas whereas commercial ranching aiming production meat for market. The positive impact result from a reduction of completion from the emerging seedlings that reduce surface fire while facilitating dominant tree growth while the negative impact comes from the creation of favorable conditions for growth and expansion of grazing-resistant grass and shrubs which leads to higher fire incidence and ecosystem structural change[5]. There is no evidence on impacts of livestock holding on dryland forest in the study area thus conducting this study were significant not only for documentation of information but also it gives clue for policymakers to enact impact full rule and regulation to ensure sustainable dryland forest management.

More than half of wood produced in the world utilized for energy production two third household in Africa use wood fuel as main sources of energy for cooking and heating according to the report of food and agriculture organization[6]. Dryland forest covers more than 40 % of world tropical forest [7]. In sub-Sharan African countries households depend on dryland forest for wood fuel, construction materials as well as industries input. Biomass dependency is extremely higher in sub-Saharan African countries for energy consumption. Dryland forest plays an important role in supporting local livelihoods but its role barely considered by responsible stakeholders in sub-Saharan African countries [8]. Firewood and charcoal production in urban areas of the continent play a significant role in supporting the livelihoods of local communities in many African countries and the same is true for Ethiopia for instance participation in charcoal business creates job opportunities for both urban and rural inhabitants in some cases more than 60 percent of household income derived from charcoal business engagement [8]. About 85 per cent of household energy derived from biomass energy sources in Ethiopia for instance from the total energy sources about 70% derived from fuelwood, 8 % from dung, 7% from agricultural residues and other sources cover the rest portion [7]. Study on dryland forest dependency conducted in southern Ethiopia shows that wood fuel extraction is the second most important sources of income for households [9]. Dryland forest defined by many scholars as types of vegetation which includes wood plats with canopy cover greater than 10 % of the land located in the dry climatic zone of the world characterized by sort rainy seasons and high temperature with higher evaporation rate [10].

Dryland and woodlands are ultimate sources of fodder for livestock which provide minerals vitamins and crude fiber for animals that makes livestock production more attractive and profitable in the area. Dryland forest also has an important role in maintaining environmental health by reducing soil erosion, nutrient recycling and carbon sequestration. The role of dryland forest is not limited to the protection of the environment but also it is a source of a variety of non-timber forest products such as honey, wild edible fruits gum and arsine and numerous medicinal plants which sustain the lives of local communities [7]. Native dryland and woodland vegetation were degraded due to unlimited grazing by domestic livestock which causes a reduction in ecological functions of dryland and woodland ecosystems [4] livestock affect woodland natural regeneration not only through removing important plant species but also by damaging the seedling physically while gathered together in a specific place for a long period of time. Intensive grazing by domestic livestock can alter the organization of vegetation by prohibiting the growth of trees and shrubs into grassland in some case the opposite is true the emergence of unpalatable species in grassland area can promote the structural conversion from grassland to woodland vegetation [4]. Grazing disturbance can lead to loss of particular plant species which may have a significant role in sustaining ecological resilient functional role in maintaining. Grazing also contributes to a succession of particular species such as annual herbs and grass which can cope with high disturbance [11]. A number of factors contribute to the degradation of dryland forest and woodland in the study area among those wood fuel productions considered as a major cause of dryland forest degradation. Generally, causes of dryland forest degradation can be classified into different major categories such as natural factors which embrace environmental and geographical factors, climatic factors and biological factors. Human-induced factors also impose pressure on dryland forest as it interacts with the environment to sustain their life [12]. Deforestation for expansion of agricultural land, deliberate burring of the forest to produce grass for their livestock, wood fuel production to generate income and resettlement are among the major human-induced causes of dryland forest degradation[12]. Overexploitation of the resources which is caused by unplanned harvesting leads to wastage of wood resources. Introduction and expansion of invasive species result in reduction and extinction of native dryland forest species for example in the study area *Prosopis juliflora* species invasive species which causes several problems but it can be used for charcoal production as well as fodder source[13]. The other categories of dryland forest degradation is that socio-economic and policy-related factors[7]. Socioeconomic factors include population growth, poverty, decreasing of living standard of households increase forest dependence hence dryland forest degradation. Lack of dryland forest management as well as the absence of land use plane, policy and rule and regulation also intensify dryland forest degradation[14].

2. Materials and method

In this study, both primary and secondary data were used. the household survey was conducted to collect primary data along with field observation were combined with secondary data collected from governmental and non-governmental reports as well as previous research work in the area.

2.1 The study area

This study was conducted around Dire Dawa town within three cluster administration zone specifically within eight kebeles called Jeldesa, Debele, Melkakero, Legedin, Kortu, Kalcha, Hulahulu, and aseliso, (see figure 1.) Dawa town is situated between 9°27'- 9°49' North latitude and 41°38'- 42°19' East longitude and found in the eastern part of the country 445 Km far from the capital city of the country; covering an area of 128802 ha The study area dominated by acacia woodland, bushes, shrublands and grassland vegetation cover as well as junipers trees in the upper side of the mountains[15].

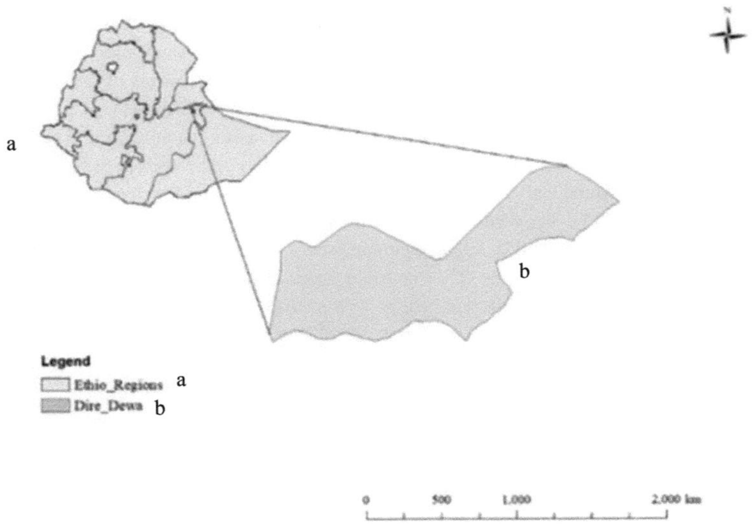

Figure 1: Map of the study area

2.2 Data collection and sampling techniques

The study area characterized by pastoralist and agro-pastoralist livelihood of which households settled far apart with irregular settlement in order to take excepted sample size of households snowball sampling techniques were used this method is highly advised by scholars to take samples from hidden population which involve illegal or disclosed population due to morality reason in our case charcoal production from natural forest is illegal activity in the country so that households may not have the willingness to participate in the survey if other methods of sampling techniques were used. On the other hand, the targeted population is not stable in specific place due to their movement for searching fodder and water for their livestock. It was difficult to estimate the total population size of targeted households to use another sampling method once gain utilization of snowball sampling was the only solution to overcome those obstacles. Snowball sampling also called referral sampling method [16]. There are about 36 peasant associations in the district of which 8 peasant association were selected based on their potential of wood fuel production. 20 households from wood fuel producer and 20 households from non-wood fuelwood producers were decided to be surveyed from each peasant association according to the available budget and manpower a total of 320 households were surveyed.

2.3 Model specification and data analysis

To analyze the collected data SPSS software version 20 were used particularly binary logit model and descriptive statistics was the major tool used to test factors which were expected to have an influence on household participation in wood fuel production from acacia woodland forest in the study area. The results of the analysis were presented in tabular as well as in diagram forms while discussed narratively. Binary logit model approximates the relationship between the dependent and independent variable which always assigned qualitative response the model specifies a functional relation between the probability of participating in the production of wood fuel and various explanatory variables. Independent variables that affect households to participate in the production of wood fuel can be expressed both qualitatively and quantitatively. Logistic regression analysis studies the association between a categorical dependent variable and a set of independent (explanatory) variables. The name logistic regression is used when the dependent variable has only two values, such as 0 and 1 or Yes and No. [17]. The dependent variable is the participation of households in wood fuel production which were coded 1 for participants in wood fuel production while 0 for non-wood fuel producers.

Drawing upon [18, 19] the logit distribution function for the participation in wood fuel production activities are specified as Logistic regression is models which construct the best fit line that describe the relationship both in terms of the direction of the relationship and estimated values between dependent variables and independent variables. Y is a dependent variable which is participation in wood fuel production (WFP) and coded 1 for HH who participate in wood fuel production from natural dryland forest whereas 0 is coded for non-participants. Age of household heads (AGEHHH), household family size (HHFS), livestock holding of households in total livestock unit (TLH_TLU) and size of arable land in hectare (SALHH) are continues independent variables will be used for analysis whereas sex of household head (SEXHHH), education status (EDUSHHH) are categorical independent variables were used for analysis.

$$WFP = \beta0 + \beta1SE + \beta2AG + \beta3FS + \beta4ES + \beta5LS + \beta6LU + \varepsilon \quad (1)$$

Where

WFP = participation in wood fuel production

SE = sex of household head (SE)

AG = Age of household head (AG)

FM = household Family size (FS)

ES = Education status of household head (ES)

LU =Livestock holding in TLU (TLH_TLU)

LS = Size of arable land in ha (SALHH)

ε = Error term

β_0= is the coefficient of the constant term

$\beta_1 - \beta_6$ = unknown parameters

Descriptive statistics also used to describe characteristics of respondents in this study mainly mean, frequency, percentile and standard deviation were used to identify major factors which influence households' participation in wood fuel production. Several tests were carried out according to the property of variables such as t-test and chi-square test.

3. Result and discussion

Before conducting an analysis of the collected data number of test for goodness of fit were conducted like Hosmer and Lemeshow test shows that the model sufficiently fit with data. A number of factors affect the household decision to participate in wood fuel production. Correlation test also conducted to see if multicollinearity exists between independent variables and the output indicates the absence of a strong correlation among independent variables.

3.1 General characteristics of respondents

Households in the study area characterized mainly by pastoralist and agro-pastoralist livelihood but they participate in wood fuel production to generate income by selling it the market for nearby urban resident and along the high way from Dire Dawa to Addis Ababa which is the capital city of the country. The general summary of households presented in Table 1 the dwellers in the area use this market opportunity to sustain their life especially during drought and famine.

Table 1: Summary description of households

Characteristics	Minimum	Maximum	Mean	Std. Deviation
Household family size	2	11	5.82	2.071
Age of household head	19	70	36.85	11.297
Size of total arable land in ha	.00	4.00	.6247	.58053
Livestock holding in TLU	0	35	8.26	7.130
Distance from forest in Km	.00	10.00	4.0075	3.01907
Years of residence in the village	6	70	36.08	11.541

(Source own survey data 2018)

From my observation, the production of wood fuel from natural acacia species is now becoming a regular job even if it not an illegal activity. Some earlier documents indicate that dry woodland area covers more than 3 million hectares which constitute important gum and resin bearing species of trees such as Boswellia, Commiphora and Striculia*[20]*. The maximum number of persons living in one household was 11 and the minimum number of persons living in one household was about 2 as well as the average family size were about 6 persons per households.

The youngest household head from the surveyed households was at the age of 19 years old and the oldest household head was at the age of 70 years old whereas the mean of household heads age was about 37 years old. The minimum arable land holding HHs (households) were 0 ha which is there were households who do not have arable land at all while the size of maximum was 4 ha of land the average size of arable land size was about 0.6 ha. The maximum distance between the household to the forest is about 10 Km. whereas the shortest distance is 0 km. which point out that HHs are living in the forest. The average distance between the household and forest area is about 4.0 km. The maximum year of residence in the area was about 70 years while the minimum years of residence were 6 years and the average years of residence in the area were about 36 years.

Livestock holding is the major factor planned to study in-depth and investigates its impact on wood fuel production. These variable measures total livestock holdings of households in tropical Livestock unit. This variable also continues variable but different animals have their own internationally agreed value to convert into tropical livestock unit and to sum up all kinds animals owned by households to change it into a single number of which is called total livestock unit. As shown in the summary table the maximum total livestock holding in tropical livestock unit was 35 and the minimum was 0 while the mean was about 8.06 TLU.

From the total households surveyed 34.4 % were female and the rest 65.6 were male-headed households. About 73.1 % of respondents were illiterate whereas only 26.9 % of respondents were literate. From this result, it can easily understand that the shortage of educational access in the study area.

Table 1: Summary of characteristics of households

Characteristics		***Frequency***	***Percent***	***Valid Percent***
Marital status	single	80	25.0	25.0
	married	240	75.0	75.0
Education status	Illiterate	234	73.1	73.1
	Literate	86	26.9	26.9
Group membership	no	280	87.5	87.5
	yes	40	12.5	12.5

(Source own survey data 2018)

More than 87 % of respondents were a member of neither formal institutions nor informal institution whereas only 12.5 percentages of respondents were a member of formal and informal institutions

3.2 Effect of livestock holding on wood fuel production

Socio-economic characteristics of the community significantly influence Forest dependency of the community. Animal husbandry is the most common livelihood type in dryland forest and woodland area dwellers. Dryland forest protection and conservation practice can provide several for socio-economic benefits for rural communities but it should be considered with households' socio-economic characteristics. Wood fuel production is mainly affected by different socio-economic factors such as household family size, sex, and age of household heads which has a great impact on the decision of household to produce wood fuel from dryland forest. The household holding of arable land and livestock has a significant ($p < 0.05$) impact on the decision of households' participation in wood fuel production in this study. Livestock holding has a negative impact which is an indicator that households allocate for time for attending livestock than participating in other economic activities.

Table 2: Impacts of livestock holding on wood fuel production

Dependent variable wood fuel production (Yes=1, No=0)	***B***	***S.E.***	***Wald***	***df***	***Sig.***	***Exp(B)***	***95% C.I. for EXP(B)***	
							Lower	***Upper***
Family size	.069	.068	1.043	1	.307	1.072	.938	1.225
Sex of household head	-.196	.275	.509	1	.475	.822	.479	1.409
Age of household head	.000	.013	.000	1	.998	1.000	.976	1.025
Size of arable land	-.973	.268	13.190	1	.000	.378	.224	.639
Livestock holding (TLU)	-.118	.021	30.429	1	.000	.889	.852	.927
Education status	-.010	.285	.001	1	.973	.991	.566	1.733
Constant	1.195	.549	4.733	1	.030	3.304		

(Source own survey data 2018)

An increase of one unit in livestock holding leads to the probability of participation in wood fuel production of households decreased by 11.1 % while the land size has a negative impact on the participation of wood fuel production. When land size increased by 1 hectare the probability of participating in wood fuel production will be decreased by 62.2 %. This result was consistent with the result found in the literature mentioned above due to the livelihood difference from those works as it was indicated the communities were pastoralist and agro-pastoralist livelihood which characterized by unstable settlement household who conquer large land size would tend to participate in wood fuel production than households with a small tract of land.

Household family size has a positive correlation with wood fuel production even if it's not statistically significant. Size of arable land and livestock holding has a negative correlation with wood fuel production which affects significantly wood fuel production. Livestock holding has expected to have a negative impact on participation in wood fuel production and the mean of wood fuel producers' livestock holding were about 5.84 TLU and 10.67 for non- wood fuel producer which indicates the negative correlation of livestock holding and wood fuel production.

Table 3: Livestock holding impacts on wood fuel production

participation in wood fuel production (n=320)		***Mean***	***Std. Deviation***	***t-value***	***Sig.***
Livestock holding in TLU	non -wood fuel producer	10.67	7.959	6.435	.000
	wood fuel producer	5.84	5.188		

(Source own survey data 2018)

Even if grazing by domestics livestock in dry woodlands can have a negative impact on the dry woodland ecosystem function by affecting its regeneration capacity on the other hand livestock holding can have a positive impact by reducing pressure on the dry woodland from commercial wood fuel production [21]. The impact of grazing is different based on the type of agroecology and type of animal and intensity of grazing ranging from destruction to the change in the ecosystem in dry woodland grazing can change it into grassland. The mean of livestock holding in tropical livestock unit between wood fuel producers and non-wood fuel producers vary significantly. The mean of livestock holding for wood fuel producers was about 5.84 TLU while 10.67 for wood fuel producers.

Huge portions of dryland population relay on animal husbandry and subsistence crop production for their livelihood. In most case livestock production combined with crop production in agro-pastoral systems. Dry lands forests are main sources of wood fuel for the rural and urban community in Ethiopia. Wood fuel production and marketing are major sources of household income in dry land areas. Extraction of wood fuel from dryland forest has an adverse impact on ecosystem services [1]. Unsustainable wood fuel production and extraction of wood form dryland forest impose a number of negative impact on the productivity of dry wood land ecosystem which leads to a reduction of service provision and supportive function of the forest. Dry land areas support the lives of more than two billion people which are more than 40% of the world's population [2]. Dryland ecosystem provides numerous services and goods such as fodder for livestock of which make it suitable for animal production in the region. Livestock resources are ultimate sources of food (meat, dairy products) and clothing raw materials such as wool and leather. The dryland areas also major sources of crop production such as wheat, barley, millet and sorghum are not only produced but also originated form dryland agro-ecological zone of the world [2]. Dry land forest area ultimate sources of fuel wood for the local community as well as nearby urban areas. Increased in number of livestock grazing in dryland forest can cause change in structure of vegetation while reducing biodiversity accompanied by increasing soil erosion but the extent of occurrence of those conditions depends on the pattern and intensity of grazing thus managed grazing plan with appropriate soil and water conservation measure will reduce human-induced negative impacts on dryland [2]. Households' dependency on wood fuel production and marketing for their livelihood affected by a number of factors for example absence of educating arable land for crop production, low level of educational attainment, large family size and absence of diversified livelihood strategies [3].

Char 1 major Objectives of wood fuel production

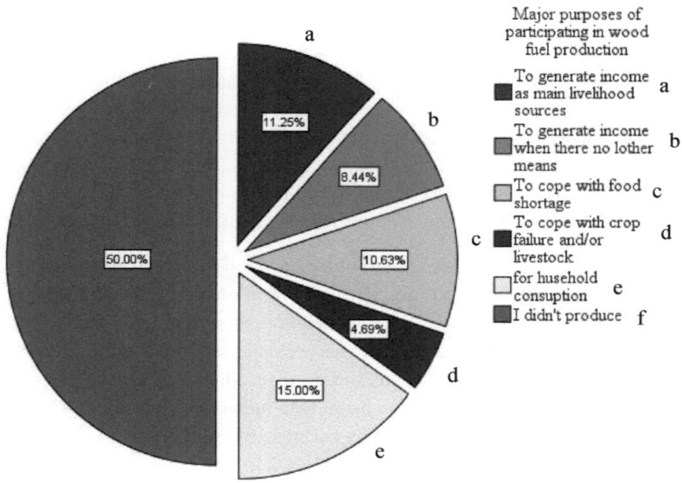

Figure 2: Major objectives of household participation in wood fuel production

In dryland area variability of rainfall is very high which induce the risk of agricultural productivity loss but dryland dwellers are have developed a variety of risk spreading strategies which include mobility diversification of income sources such as wood fuel productions and engagement in non-agricultural works[3]. Households in the study area households participate in wood fuel production for different purpose such as to generate income for main sources of livelihood income (11.25%), to generate income at the time of other means of income are not available (8.44%), to escape food shortage problems (10.63%), to compensate loss of income due to livestock death and/or loss of crop productivity (4.69%), to provide energy for household consumption (15%) and the rest (50%) households were non-wood fuel producer.

4. Conclusion

Ethiopia is a country implementing a green economic growth strategy to meet the vision to be a middle-income country by 2030. As mentioned in the second growth and transformation plan. In order to achieve this vision, proper management of natural resources is critical not only to conserve the existing degraded forest but also to improve the livelihood of the communities. About 75 % of Ethiopia's land mass considered as dry land which is characterized by a shortage of rainfall and high temperature according to the report of[11, 22]. Dryland forest and woodland is contributing for the huge part of the existed forest in Ethiopia and expected to reach 55 million heaters which are about 48.6 % of total landmass[23] Forest degradation in Ethiopia are still increasing in alarming rate as estimated by[24] two million ha of forest lost every year. This loss of forest creates interrelated environmental, social and economic problems which cause a low level of living standards and affects economic growth as a whole. Woodland resources of the country are decreasing from time to time due to increasing demand for additional arable and pasture land as well as increasing wood fuel demand associated with increasing population growth[25]

About 60 % of the total land area of Ethiopia is covered by acacia dominated dry woodland which accounts most portion of wood fuel production for domestic consumption the country meets its energy needs which is more than 96 % comes from wood fuel according to the report of[26]. To make an appropriate intervention to solve the problems in wood fuel production sector, it is critical to identify major factors affecting participation in wood fuel production. This study shows that the education status of the community is a major determinant in wood fuel production. More educated people can cope up with risk and shocks by participating in activities other than livestock production and crop production to sustain their life during hardship so investing on the expansion of formal and informal education will increase the probability of participating in wood fuel production. On the other hand females, headed households are fewer participants in wood fuel production to enhance fair and equitable utilization of natural resources female-headed households should get special assistance. The result indicates that there is a chance of specialization in the production of wood fuel which increases income for the country if it is managed properly and utilized suitably. Increased in number of livestock grazing in dryland forest can cause change in structure of vegetation while reducing biodiversity accompanied by increasing soil erosion but the extent of occurrence of those conditions depends on the pattern and intensity of grazing thus managed grazing plan with appropriate soil and water conservation measure will reduce human induced negative impacts on dryland[27].

References

1. Maass, J.M., et al., *Ecosystem services of tropical dry forests: insights from longterm ecological and social research on the Pacific Coast of Mexico.* Ecology and society: a journal of integrative science for resilience and sustainability, 2005. **10**(1): p. 1-23.
2. Weber, K.T. and S. Horst, *Desertification and livestock grazing: The roles of sedentarization, mobility and rest.* Pastoralism: Research, Policy and Practice, 2011. **1**(1): p. 19.
3. Lawry, S., R. McLain, and H. Kassa, *Strengthening the resiliency of dryland forest-based livelihoods in Ethiopia and South Sudan: a review of literature on the interaction between dryland forests, livelihoods and forest governance.* Vol. 182. 2015: CIFOR.
4. Raymond H., P., Neil E.,1 Froend, & Ladd, Phillip G., *Grazing in remnant woodland vegetation: changes in species composition and life form groups,* in *Opulus Press Uppsala.* 1995, Vegetation Science Sweden. p. 121-130,.
5. Belsky, A.J. and D.M. Blumenthal, *Effects of Livestock Grazing on Stand Dynamics and Soils in Upland Forests of the Interior West: Efectos del Pastoreo sobre la Dinámica de Árboles y Suelos en Bosques en el Altiplano del Occidente Interior.* Conservation Biology, 1997. **11**(2): p. 315-327.
6. Van Dam, J., et al., *The charcoal transition: greening the charcoal value chain to mitigate climate change and improve local livelihoods.* 2017.
7. Teketay, D., *Causes and consequences of dryland forest degradation in Sub-Saharan Africa.* Vol. 24. 2004. 3-20.
8. Chidumayo, E.N. and D.J. Gumbo, *The dry forests and woodlands of Africa: managing for products and services.* 2010: Earthscan.
9. Fikir, D., W. Tadesse, and A. Gure, *Economic contribution to local livelihoods and households dependency on dry land forest products in Hammer District, Southeastern Ethiopia.* International Journal of Forestry Research, 2016. **2016**.
10. Ndegwa, G.M., *Evaluating Dry Woodlands Degradation and On-farm Tree Management in Kenyan Drylands.* 2017, Universität Passau.
11. Giorgis, K. *Agricultural and pastoral technologies and practices for climate change adaptation in dryland areas of Ethiopia.* in *proceeding of a high level policy forum workshop.* 2014. Addis Ababa: Afar National Regional State, and UNDP Country Office.
12. Kiruki, H.M., et al., *The effect of charcoal production and other land uses on diversity, structure and regeneration of woodlands in a semi-arid area in Kenya.* Forest Ecology and Management, 2017. **391**: p. 282-295.
13. Berhanu, A.a.T., Getachew, *The Prosopis dilemma, impacts on dryland biodiversity and some controlling method.* Journal of the Drylands, 2006. **1**(2): p. 158--164.
14. COOKE, P., KÖHLIN, G., & HYDE, W. , *Fuel wood, forests and community management – evidence from household studies.* Environment and Development Economics, 2008. **13(1)**(S1355770X0700397X): p. 103-135.
15. Abay, N.G., *Assessment on Acacia Woodland Degradation in Dire Dawa* International Journal of Regional Development, 2018.
16. Dragan, I.-M. and A. Isaic-Maniu, *Snowball sampling completion.* Journal of Studies in Social Sciences, 2013. **5**(2).
17. Gujarati, *Basics of Econometrics* 1988, New York.

18. Gujarati, D.N. and D.C. Porter, *Essentials of econometrics*. Vol. 2. 1999: Irwin/McGraw-Hill Singapore.
19. Aldrich, J.H., et al., *Linear probability, logit, and probit models* 1984.
20. Mendum, R. and M. Njenga, *Integrating wood fuels into agriculture and food security agendas and research in sub-Saharan Africa*. FACETS, 2018. **3**(1): p. 1-11.
21. Pettit, N.E.a.F., Raymond H and Ladd, Phillip G, *Grazing in remnant woodland vegetation: changes in species composition and life form groups*. Journal of Vegetation Science, 1995. **6**(1): p. 121--130.
22. Adhikari, B., S. Di Falco, and J.C. Lovett, *Household characteristics and forest dependency: evidence from common property forest management in Nepal*. Ecological economics, 2004. **48**(2): p. 245-257.
23. MoARD, *Forest Resources of Ethiopia*. . 2004: Addis Ababa, Ethiopia.
24. FAO, *Global Forest Resource Assessment. Main Report FA Forestry Paper*. 2010, Food and Agricultural Organization United Nations, : Rome.
25. Reusing, M., *Mapping woodland vegetation in SW Ethiopia based on a hybrid model*. 2017, International Journal of Sustainable Development & World Ecology,. p. 251-259.
26. Gebremedhin, T., *Biomass estimation of herbaceous and woody vegetation in closed areas of Northern Ethiopia*. 2002.
27. White, R.P. and J. Nackoney, *Drylands, people, and ecosystem goods and services: a web-based geospatial analysis (PDF version)*. World Resources Institute ([Available at: http://pdf. wri. org/drylands. pdf accessed on 30/01/2012]), 2003.